W9-CXU-263

Make It! Ship It!

by Janine Scott

Content and Reading Adviser: Mary Beth Fletcher, Ed.D.
Educational Consultant/Reading Specialist
The Carroll School, Lincoln, Massachusetts

COMPASS POINT BOOKS
Minneapolis, Minnesota

Compass Point Books
3722 West 50th Street, #115
Minneapolis, MN 55410

Visit Compass Point Books on the Internet at *www.compasspointbooks.com*
or e-mail your request to *custserv@compasspointbooks.com*

Photographs ©: EyeWire/Getty Images, cover, 12; TRIP, 5, 8, 13; Gamma/Kurita Kaku, 6; Bob
Rowan/Corbis, 7; Gamma/P. Vasquez-Cunningham, 9; Imagestate, 10, 11, 17; TRIP/H. Rogers, 14, 21;
PhotoDisc/Getty Images, 15, 19; TRIP/A. Dalton, 16; TRIP/M. Shirley, 18; TRIP/Eric Smith, 20.

Project Manager: Rebecca Weber McEwen
Editor: Heidi Schoof
Photo Researcher: Image Select International Limited
Photo Selectors: Rebecca Weber McEwen and Heidi Schoof
Designer: Jaime Martens

Library of Congress Cataloging-in-Publication Data

Scott, Janine.
  Make it! ship it! / by Janine Scott.
    p. cm. — (Spyglass books)
Summary: Briefly introduces how goods, from paperclips to pick-up
trucks, are manufactured and distributed around the world.
Includes bibliographical references and index.
  ISBN 0-7565-0363-9
1. Manufacturing industries—Juvenile literature. 2. Manufacturing
processes—Juvenile literature. 3. Physical distribution of goods—
Juvenile literature. [1. Manufacturing industries. 2. Manufacturing
processes. 3. Physical distribution of goods.] I. Title.
II. Series.
  HD9720.5 .S26 2002
  338.4'767—dc21
                            2002002738

Contents

Making Things

All around the world, people are making things, from basketballs to big blue cars. It sometimes takes many different steps for these things to be made.

What Will They Make?

The first step is deciding what needs to be made. A *designer* figures out how to make this new thing work. It should be safe and easy to use.

Did You Know?

Car companies have new car designs every year.

What Is It Made Of?

Next, people need to find the *raw materials* to make this new thing. For example, tires are made of rubber. Paper is made of wood. Rugs are made of sheep's wool.

Did You Know?
Rubber comes from trees.

How Is It Made?

The new thing might be made by changing the raw materials into something new. For example, olives are made into olive oil.

Olives

Olive oil

Did You Know?

Someday this wheat might be made into breakfast cereal.

11

Handy Machines

New things are usually made or built in large *factories.* People use machines and other tools to do work that is too hard or dangerous for people to do themselves.

Did You Know?

Crash test dummies are used to make sure cars are safe.

People Power

Factories need people to build the new thing. Other people make sure the new thing works. Some pack the thing into boxes. Others help sell the thing to people who need it.

Did You Know?
Companies often build factories near big cities, to be close to the people who will work there.

Ship It!

Once the new thing is made, it is time to ship it. Big ships carry things across oceans to other places. Smaller boats carry things across lakes and rivers.

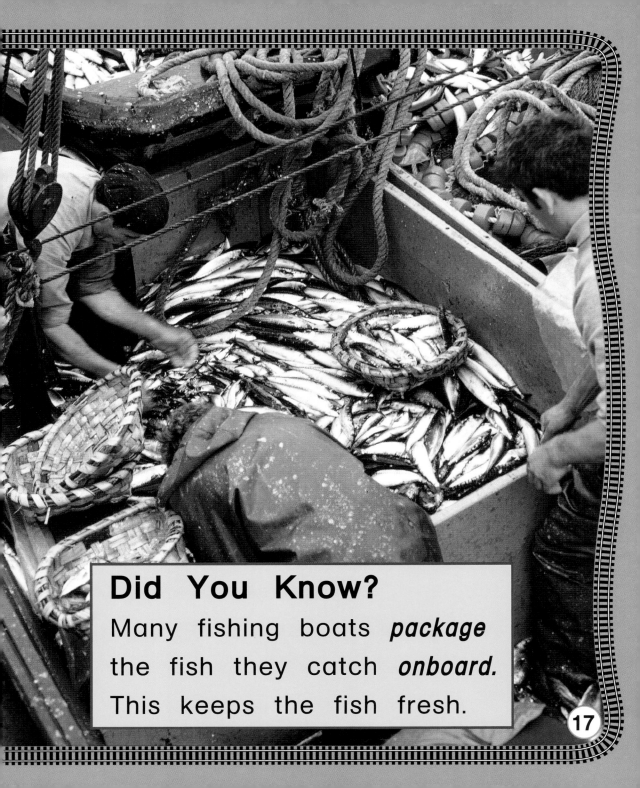

Did You Know?

Many fishing boats *package* the fish they catch *onboard.* This keeps the fish fresh.

17

Fly It!

When things such as food and flowers must be kept fresh, flying is the fastest way to get them where they need to go.

Did You Know?

It is expensive to ship things by air. Airplanes do not have a lot of space, and they use costly *fuel.*

Truck It!

Sometimes people ship things over land. Trains can carry heavy loads a long way.

Some trucks carry things between cities.

Did You Know?

In some countries, people use animals to carry heavy things.

Glossary

designer–a person who comes up with ideas for new things and decides how they will be made

factory–a place where people and machines work to make things

fuel–something that makes heat or energy, such as oil

onboard–when something is on a ship, train, or airplane

package–to wrap something up

raw materials–things found in nature that people can use or make into something new

Learn More

Books

Saunders-Smith, Gail. *Trucks.* Mankato, Minn.: Pebble Books, 1998.

Stone, Lynn M. *Freight Trains.* Vero Beach, Fla.: The Rourke Book Company, 1999.

Sullivan Hill, Lee. *Get Around with Cargo.* Minneapolis, Minn.: Carolrhoda Books, 1999.

Web Site

www.brainpop.com
(click on "Assembly Line")

Index

GR: I
Word Count: 243

From Janine Scott

I live in New Zealand, and have
two daughters. They love to read
books that are full of fun facts
and features. I hope you do, too!